BUILDING YOUR OWN BASIC GO KART TIPS AND TRICKS

THOMAS BRESSON

Illustrations by Oriane Michel

Layout and Photographs by Thomas Bresson

Schema on page 1 by Azusa Parts

The author assumes no responsibility
for any injury, loss, or damage caused or sustained
as a consequence of the application of the book's contents

Contents

Introduction

CHAPTER 1	Kart Information	1
CHAPTER 2	Painting and Necessary Supplies	5
CHAPTER 3	Installing Spindles and Front Wheels	9
CHAPTER 4	Steering Assembly	13
CHAPTER 5	Installing Axle Bearings	19
CHAPTER 6	Adding and Installing Brake, Sprocket, and Hub	23
CHAPTER 7	Assembling and Mounting Rear Wheels	27

CONTENTS

CHAPTER 8	Installing Pedals and Brake Rod	31
CHAPTER 9	Mounting Engine and Clutch	35
CHAPTER 10	Chain and Rear Sprocket Installation	39
CHAPTER 11	Bucket Seat, Axle Cover, and Throttle Rod	43
CHAPTER 12	Tools and Supplies Check-List	49

Acknowledgements

Introduction

My name is Thomas Bresson and I will be helping you in your journey of building a kart.

I am a teenager in High School and an avid motorsports fan. I have always had a passion for racing and for speed, from little toy cars and remote-controlled cars to karting competitively. I love watching Formula 1, NASCAR, IndyCar, and endurance racing.

The kart that is built in this book is intended for recreational activities, not for competitive racing.

INTRODUCTION

If you need any information regarding competitive karts, you can contact me and I will do my best to help you!

Standard tools are usually enough to build the kart, but it is possible to run into complications that require additional tools listed in each chapter.

Using my own experience, I will be guiding you to avoid running into the same difficulties as I did. I am very excited to help you with building your kart and I wish you the best of luck!

CHAPTER 1

Kart Information

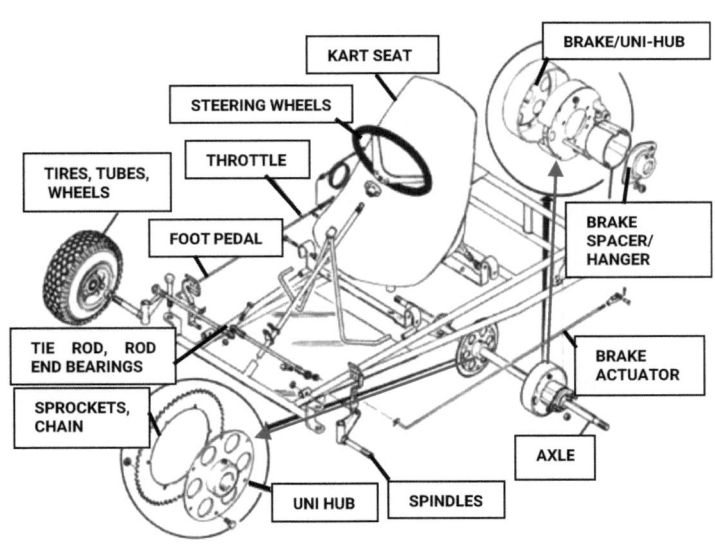

GO KART PARTS LOCATOR

As previously mentioned, this type of kart is intended for recreational purposes and not for racing. I use this kart to go around my neighborhood, to access areas near my house, and to tow friends in a trailer.

This kart is a generally simple machine and does not require much maintenance or time. Something to take note of is that this kart requires a clutch, which does not come with the engine. As a broad overview, for this kart, you will need a chassis, rims, tires, spindles, steering components, a brake drum, throttle and brake pedals, axles and bearings, a throttle connection kit, a chain, an engine, a sprocket and hub, and a seat.

To find all the necessary parts, you can go online for the individual parts or buy a kit. Most kits don't include a chassis but I would recommend buying a chassis that does not require welding. This kart can easily be assembled at home.

I recommend having about 12 x 8 feet of space to be able to organize all the components. Having some support to put the chassis on is also very helpful.

Once you have all the parts, organizing them by each step in various boxes would be the simplest way to go about keeping a clean and tidy building area.

CHAPTER 2

Painting and Necessary Supplies

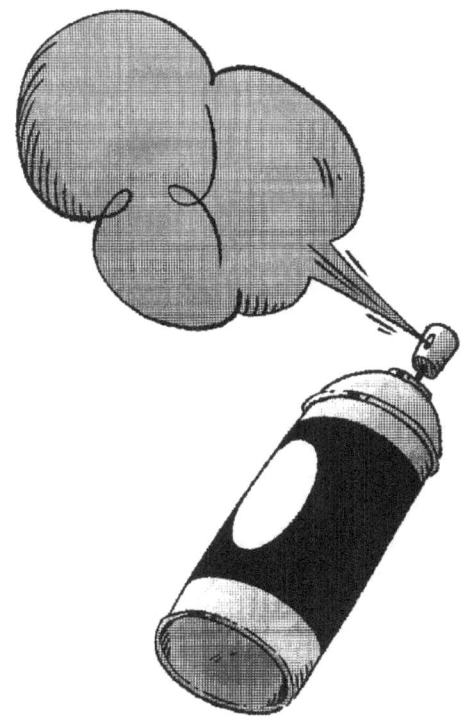

The first step in the actual construction of the kart is to get all the necessary painting supplies. The easiest way to paint this type of kart is by *spray painting*. To be safe, buy 2 cans of primer (15 oz. each) and 4-5 cans of spray paint (12 oz. each). You can have multiple colors on the same part or keep each part the same color.

It is also helpful to have a tarp, blue painter's tape, and a cloth. I chose to paint the wheel rims, the steering hoop, the steering rod, and the pedals.

Spray painting the kart components with primer

Start the process by applying 3 coats of primer, waiting 4 to 6 hours between each coat. After each coat of primer is dry, make sure to sand off any bumps or irregularities to keep a completely smooth surface.

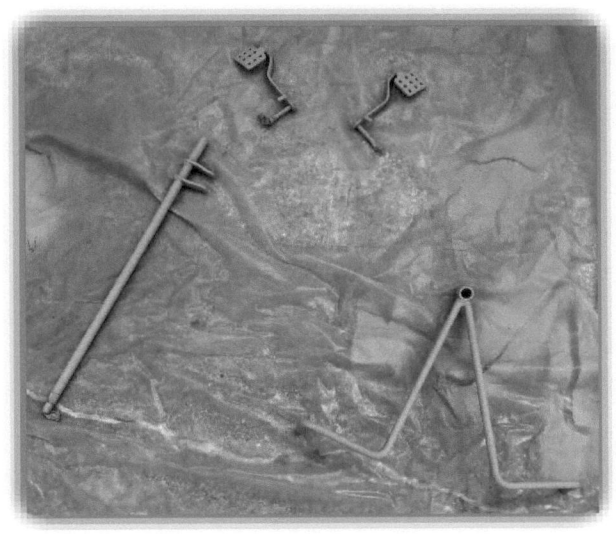

Applying primer on the pedals, steering shaft, and steering hoop

Then add 3 or 4 coats of paint and again wait 4 to 6 hours in between.

While painting, do not bring the can too close to the part you are painting, and make sure to constantly move the can to spread the paint evenly and prevent dripping.

If too much paint is applied, grab a cloth and tap on the part to remove the excess paint.

If the paint dries up and the surface is not smooth, you can sand it and add another layer of paint. This process takes several cycles as you must rotate the part to paint the front, back, and sides of it. While painting, make sure to cover up any areas where bolts will be screwed or where it will be in contact with other parts.

Useful supplies for this step: tarp, blue painter's tape, white cloth, 2 cans of primer, 4-5 cans of spray paint

CHAPTER 3

Installing Spindles and Front Wheels

Installing the *spindles* is pretty straightforward. The only small detail to remember is adding the Nyliner split bushings to protect the chassis and spindle when the kart is in use. Once everything is installed, make sure to tighten the bolts and locknuts.

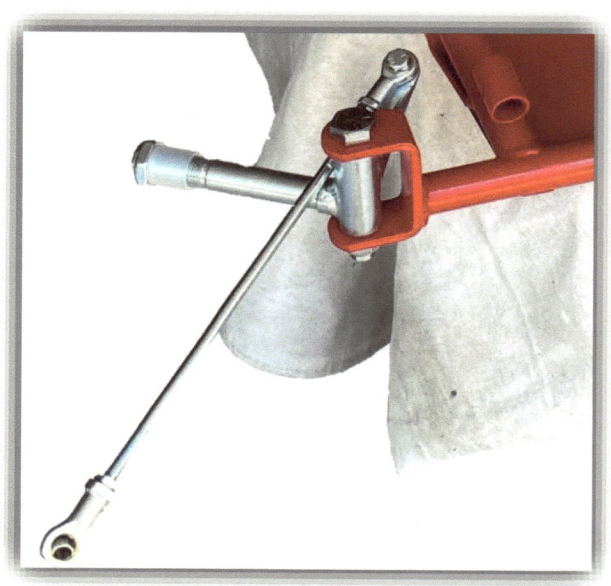

Example of a completed spindle installation

To assemble the *front wheels*, take one of the half rims and place the inner tube around that rim. The half rims for the front wheels should each contain a bearing.

Then line up the valve stem with the hole in the rim and place the other half of the rim on top of it. Ensure the inner tube does not get pinched between the wheel halves and then add and tighten the bolts and locknuts.

Fully built front wheels

Once the two rims are fully tightened, pump the tires to the recommended psi, usually specified on the tire.

Useful tools for this step: bike pump with psi reader

CHAPTER 4

Steering Assembly

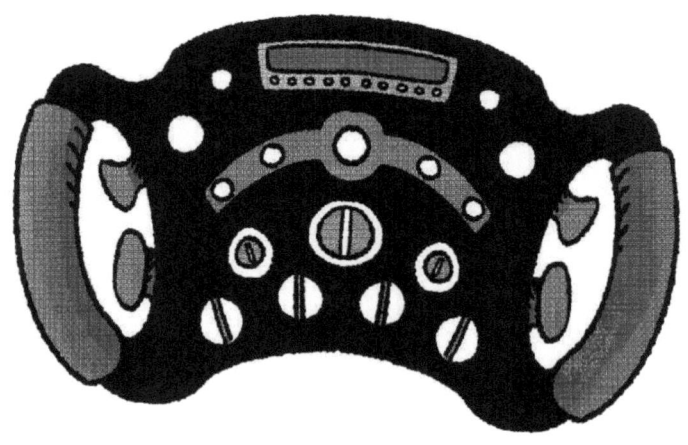

For my kart, I had to assemble the *steering shaft and hoop* by first inserting a shaft collar at the end of the shaft and then adding a Nyliner split bushing or something similar to protect the shaft.

Assembly of steering and hoop with tie rods

Then I placed the shaft into a hole in the chassis but did not attach or tighten anything for the time being. Next, I inserted the second shaft collar.

This step may require some sanding of the paint if the shaft collar does not slide on easily.

Next, add an *A-shaped steering hoop* or some sort of support for the steering shaft. In my case, I had to lower the A-shaped hoop until I could put the ends into the slot bushings that are welded onto the chassis. When I tried to put the steering hoop into the slot bushings, the ends did not slide in properly as a bump of metal from welding was sticking out. I had to use a file small enough to go into the hole to file it down. I also had to use a rubber mallet to tap it in as it was difficult to slide in.

Once everything was properly installed, I added a *Nyliner split bushing* for protection. But when I tried to insert the Nyliner split bushing, it wouldn't fit in properly, so I had to cut it in half and only use half of it. If you have shaft collars, now shift them back to their proper position and tighten the set screws.

When all of this is completed, you may move on to attaching the *tie rods*. As you attach them, make sure that the tie rods are the proper length to keep the front wheels straight.

Completed Steering assembly

Then attach the *steering wheel* and ensure that it is tight. The last step is to add a *snap ring* if you have a groove somewhere in the shaft.

When I tried doing this step, I was unable to open the snap ring wide enough to put it on, so I had to go buy special pliers to finish the job.

Useful tools for this step: file, snap ring pliers, sanding paper (1000-1200 grit), rubber mallet

CHAPTER 5

Installing Axle Bearings

To attach the *bearings*, I had to sandwich each *axle bearing* between a pair of *flangettes*. Once completed, I very loosely tightened the bearing/flangette assembly to one side of the chassis.

Finished bearing/flangette assembly

On the other side, you will need to attach the brake spacer to the bearing/flangette assembly if you are using a drum brake.

You will want the bolts to sit in the grooves of the brake spacer.

Once you have loosely attached both bearing/flangette assemblies on each side of the chassis, grab the rear axle and slip it through the two axle bearings to create the right angle for the bearings.

Once the axle is in position, tighten the bearing/flangette assembly and then remove the axle.

CHAPTER 6

Adding and Installing Brake, Sprocket, and Hub

For the installation of a *drum brake*, grab the brake and slide the back of the brake into the spacer. Align the holes, insert the bolts, and make sure that the arm of the brake is pointing upwards. Once everything is lined up, install the bolts and tighten them, but do not tighten them too tightly.

The next step is to attach the *sprocket* and *uni-hub*. This is pretty straightforward and just requires you to insert bolts and tighten the hex nuts.

Drum brake before completion

Then insert the *axle* from the right side of the kart and through the right-hand bearing; do not insert the axle fully. Slide the brake drum over the brake followed by the hub assembly.

Once this is completed, slide the axle through the left bearing. Ensure the axle is centered and then tighten the bearing set screws to hold the axle in place.

Finalized drum brake assembly

Then tighten the brake locknuts and place the brake drum over the brake shoes, but back the drum off 1/16" from the brake backing plate.

Lastly, add the keys to the brake drum, and make sure to add and tighten the set screws. At this point, it is normal that the sprocket and hub assembly are floating freely.

CHAPTER 7

Assembling and Mounting Rear Wheels

To assemble the *rear wheels*, use the same process as the front wheels, but each pair of wheels should contain 2 different *half wheels* in most cases. Typically, one half-wheel should be indented whereas the other half-wheel should have a small slit to slide in a key.

Half wheels used for rear wheels assembly

Insert the *inner tubes* and make sure the stem fits into the stem notch. Bolt the two halves together and partially pump the inner tubes.

Then, put the tire over the inner tube and fully pump the wheels. Make sure that the half wheel with the hole/indent points away from the frame of the kart and finally slide the wheels onto the axle.

Rear wheels and inner tubes

When I tried to tighten the wheels to the axle, the lockout would not go on and would get stuck so I put some oil inside the nuts.

Make sure that the wheels are keyed and that they are well attached before using the kart.

Useful supplies for this step: bike oil

CHAPTER 8

Installing Pedals and Brake Rod

For this step, start by attaching the *pedals* to the kart frame while making sure that they can move freely.

On my kart, I had to put the pedals through holes in a part that was welded to the front of the frame. Once the pedals are securely attached, add a split lock washer and then tighten the locknuts. Ensure that the pedals aren't loose but are still able to move.

Fully assembled pedal without throttle attachment

Then, attach the *brake control rod* to the brake pedal. I put the end of the rod that had a 90-degree bend through a hole in the pedal and put a flat-type speed nut to hold it in place.

Next, I had to twist a clevis on the other end of the rod and attach it to the arm of the brake using a cotter pin and a clevis pin.

Attached brake control rod

If you are using any type of drum brake, you will have to connect the pedal and brake arm through some sort of stick or wire.

If you are using disk brakes, you may be able to use a rod attachment, similar to a drum brake. However, at other times, a hydraulic system is required.

CHAPTER 9

Mounting Engine and Clutch

The *engine* usually does not come with a kart kit if you buy one. There are many different engines you can get but I would recommend looking online for advice from people who have built a similar kart. If you opt for a lawn mower-style engine, you must make sure you buy one with a horizontal shaft and not a vertical shaft.

Possible engine

To mount the engine, put it on a strong flat surface and attach it by tightening it with bolts, but do not fully tighten them at first.

For the *clutch*, you must buy one that works with your engine shaft. Afterward, all there is to do is slide the clutch onto the engine shaft with the key. In my case, the key that came with my clutch was too long so I had to cut part of it off. Similarly to the engine, do not fully tighten the set screws yet.

Engine attachment onto kart frame

Useful tools: Saw for metal

CHAPTER 10

Chain and Rear Sprocket Installation

The next step is to add the *chain* to the kart, but first, you must line up the *sprocket* and *clutch* and tighten the set screws on both of them.

To put the chain on, wrap the chain around the teeth of the sprocket and clutch and mark how long you need the chain to be. Then, use a chain tool to remove the extra chain and wrap it around the clutch and sprocket.

Completed chain assembly

When attaching the two ends of the chain, make sure that the connecting link's spring clip points in the direction the chain moves.

I initially struggled with putting the chain on as I did not have the right chain tool. Once I got the chain to be the right size, every time I got the chain wrapped around the sprocket it would come off the clutch or vice versa. Eventually, I got a second person to help me keep the chain from flying off and it proved to be much easier.

Once the chain is fully wrapped, adjust the position of the engine so that the chain is reasonably tight with some margin. Once everything is lined up and set up properly fully tighten down the engine.

Useful tools for this step: chain tool

CHAPTER 11

Bucket Seat, Axle Cover, and Throttle Rod

There are many different ways to mount a seat and different types of seats to mount. In my case, I mounted a *bucket seat*.

To do so, I had to attach mounting brackets to the bottom of the frame of my kart and tighten them using bolts and nuts. I started by keeping the bolts loose so I could move the mounting brackets around to get them the right distance from each other to attach the seat.

Finalized seat + throttle attachment in process

Once I got everything set up correctly, I just had to tighten the mounting bracket bolts and tighten the bolts into the bottom of the seat.

Next, to complete the *throttle installation*, I removed the air filter from my engine that was in the way. I then attached the throttle cable to the throttle on the engine by screwing the cable to it. I connected the cable to the pedal through another rod which I attached through a hole in the pedal with another flat-type speed nut.

Cable attachment to engine

The cable went through a hole in the other end of a rod and I put a wire stop at the end of the cable.

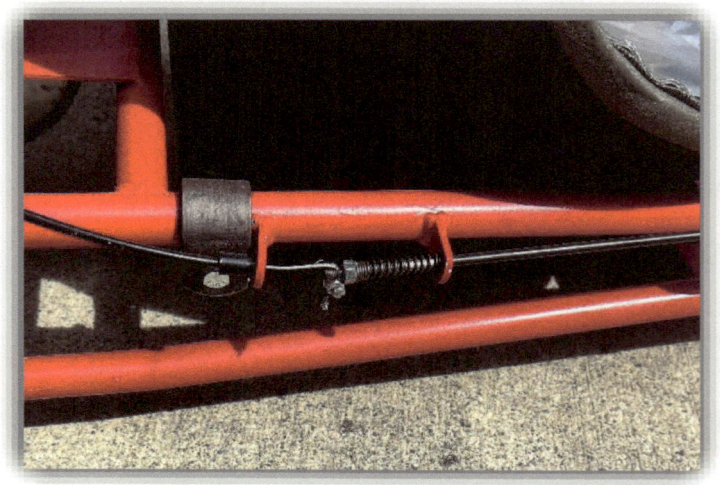

Finalized throttle connection

I also added a spring toward the end of the rod so that the pedal didn't get stuck while accelerating. I cut off any extra cable and finished by adding *axle covers* to protect the axle from any debris you could encounter.

Now that you have completed the assembly of your kart, I would suggest testing the go-kart at slow speeds to ensure that everything is functioning properly.

I hope I was able to assist you and provide some helpful tips for the construction of your kart!

Good luck and have fun!

CHAPTER 12

Useful Tools
and Supplies Check-List

Recommended Tools and Supplies:

- Tarp
- Blue Painter's Tape
- White Cloth
- 2 Cans of Primer
- 4-5 Cans of Spray Paint
- Bike Pump with psi reader
- File
- Snap Ring Pliers
- Sandpaper (1000-1200 grit)
- Rubber Mallet
- Bike Oil
- Chain Tool

Here it is!

Completed kart

Acknowledgements

I hope that this guide was able to help you in your kart-building experience! I would like to thank some key people who made it possible to build this kart, race karts, and nurture my motorsports passion.

First, I am thankful for my best friend, Luca Nowinski, who helped me reignite my passion for motorsports and bring me back into this sport. His keen eye enabled me to articulate the message efficiently for any novice in go-kart building.

Thank you also to Oriane Michel for taking the time to listen, to adapt to my whole project, and for making it lively with her illustrations.

ACKNOWLEDGEMENTS

I owe a special thanks to Sean Barber, for sharing his expertise in painting and guiding me during the first step of building my kart.

And finally, I want to thank my parents for pushing me to pursue my dreams on and off the track and to turn them into a reality.

www.ingramcontent.com/pod-product-compliance
Lightning Source LLC
Chambersburg PA
CBHW040324220526
45473CB00009B/2566